THE NEW LAWS OF PHYSICS

The solution to the major problems of theoretical
physics based on a single concept: the charge factor

Charlemagne Olivier Vernet

Ibukku is an auto-publishing company. The content of this work is the responsibility of the author and do not necessarily reflect the views of the publisher.

Published by Ibukku.
www.Ibukku.com
Graphic design: Índigo Estudio Gráfico
Copyright © 2018 Charlemagne Olivier Vernet
ISBN Paperback: 978-1-64086-217-3
ISBN eBook: 978-1-64086-218-0
Library of Congress Control Number: 2018951655

Index

Let us hope that the technology resulting from the new theory will contribute to improving the quality of life on the planet.

ACKNOWLEDGMENTS

The author thanks, in advance, the international laboratories which will be able to experimentally verify the predictions of the new theory, the theoretical physicists who will use it to explain the fundamental interactions coherently, and the astrophysicists who will use the law of quantum gravity, formulated in this book, to accurately describe the planetary orbits.

PREFACE

In this piece, phenomena as seemingly different as electro-magnetism, weak interactions, strong interactions, quantum gravity and supra-force (the unified fifth fundamental interaction), as well as the constant speed of light and the cosmological constant, are described in a cohesive way through a single concept, the notion of charge. In fact, according to this new approach, the first 5 are mediated by electrically charged messenger particles that interact with electrically charged masses, causing attractions, repulsions, disintegrations and transformations. While, the last 2 are also fundamental interactions which are mediated, not by bosons, but by messenger charges which enable them, in the case of the constant speed of light, to convert masses into energy in the form of gamma rays and, in the case of the cosmological constant, to drive 4-dimensional space-time to generate an expanding Universe.

The different approaches are carried out with mathematical consistency, in accordance with the principles of symmetry and respecting the laws of conservation; they also have solid scientific support.

CHAPTER ONE
The concept of charge

Why, if the 4 fundamental interactions of physics are basically different manifestations of the same phenomenon, are they currently described by different concepts? Undoubtedly, such a fact indicates that the various theories, although correct, are approximations of a much deeper reality and consequently are not capable of solving problems such as the calculations of strong interactions, the quantization of gravity, the unification of the 4 fundamental interactions into a supra-force, the behavior of the basic components of matter and the timing of the Big Bang.

In order to coherently explain this profound reality, it is essential to have a theory that is not approximate but fundamental, which, based on a single concept, describes the various manifestations of the fundamental interactions. On the horizon is the Fundamental Theory of Charges, which is capable of carrying out this difficult task. According to this proposal, no particle is electrically neutral, but rather, each has a particular electrical charge, which can be classified as either category 1, 2, 3 or 4. Current physics only takes into account category 2, carried by particles such as the electron and the proton, ignoring category 1, which corresponds to the neutrino and the neutron, category 3 is related to the messenger particle of strong interactions, the gluon, and category 4 belongs to high-energy photons such as ultraviolet light, gamma rays and X-rays. Such omissions prevent the scientific community from realizing that each of the fundamental interactions is related to one of the four categories of electrical charge, which means that a single concept, that of charge, is capable of explaining, with mathematical consistency and in accordance with the principles of symmetry, the various manifestations of the fundamental interactions.

The 4 categories of electrical charge

The following equations propose the equivalence between a magnetic monopole (represented by the letter D in honor of Dirac who suggested its existence) and an electrical charge of category 1; between a magnetic dipole and an electrical charge of category 2; between a dyon (electromagnetic charge) and an electrical charge of category 3; and between a high energy photon (ultraviolet light, gamma rays and X-rays) and an electrical charge of category 4. It should be noted that the value -1 included in each of the formulas is a constant (the constant -1), that to multiply by a magnetic monopole, it converts it into an electrical charge of category 1 and vice versa.

$$e = -D \text{ X } -1 \qquad (1)$$
where e is a positive electrical charge of category 1, and -D, a negative magnetic charge.

$$e^2 = (-D \text{ X } -1)^2 \qquad (2)$$
where e^2 is a positive electrical charge of category 2, and $(-D \text{ X } -1)^2$, a magnetic dipole.

$$-e^3 = (-D \text{ X } -1)(-e^2) \qquad (3)$$
where $-e^3$ is a negative electrical charge of category 3, and $(-D \text{ X } -1)(-e^2)$, a dyon (electromagnetic charge).

$$e^4 = (-D \text{ X } -1)(e^3) \qquad (4)$$
where e^4 is a positive electrical charge of category 4, and $(-D \text{ X } -1)(e^3)$, a high-energy photon.

It may seem strange to the reader to speak of a negative and, consequently, a positive magnetic charge, rather than a north and south pole. However, as will be seen below, such a notion is key to explaining, with mathematical consistency and observance of the laws of conservation, many enigmas of theoretical physics.

Equations (1-4) suggest the existence of 4 categories of electrical charge, each of which is transported by some particle, either a fermion (a half-integer spin particle) or boson (an integer spin particle).

The basic components of matter

In the sixth chapter, there will be theoretical proof that, at high energy, within an infinitely dense point, the 3 atomic components, the electron, the proton and the neutron, decay through weak interactions generating 6 fundamental leptons: the neutrino, the antineutrino, the levorotatory electron, the levorotatory positron, the dextrorotatory electron and the dextrorotatory positron. All of which, according to the fundamental theory of charge, constitute the basic components of matter, being able to combine to create quarks and messenger particles. Each of these leptons has 3 essential characteristics: a magnetic and/or electric charge, a $+1/2$ or $-1/2$ spin and a close relationship with a certain Lie group, as can be seen below:

1. Neutrino	$-D(-1)$	e	spin$+1/2$	G2
2. Antineutrino	$D(+1)$	-e	spin$+1/2$	SU(3)
3. Levorotatory electron		$-e^2(-1)$	spin$+1/2$	SU(2)
4. Levorotatory positron		$e^2(+1)$	spin$+1/2$	U(1)
5. Dextrorotatory electron		$-e^2(-1)$	spin$-1/2$	F4
6. Dextrorotatory Positron		$e^2(+1)$	spin$-1/2$	E6

Based on these 3 characteristics of the 6 fundamental leptons, and always taking into account the aforementioned infinitely dense point, the origin not only of electromagnetism, weak interactions, strong interactions, quantum gravity and supra-force, but also of the energy source from which these forces are fed will be explained below: the quantum fluctuations of the vacuum, a phenomenon also known as the Heisenberg uncertainty principle, consists of virtual pairs of particles and antiparticles being constantly created and destroyed in vacuum.

Origin of electromagnetism

The levorotatory positron U(1) is attracted by the dextrorotatory electron F4 to give rise to electromagnetism g':

$$g'=U(1) \text{ X } F4 \tag{5}$$

Origin of weak interactions:

The dextrorotatory positron E6 is attracted by the levorotatory electron SU(2) to give rise to the weak interactions g:

$$g=E6 \text{ X } SU(2) \tag{6}$$

Origin of the strong interactions gs:

The G2 neutrino is attracted by the SU(3) antineutrino to give rise to the strong interactions gs:

$$gs=G2 \text{ x } SU(3) \tag{7}$$

Origin of quantum gravity:

The electromagnetism U(1) X F4 is attracted by the weak interactions E6 X SU(2) to give rise to the quantum gravity SU(4):

$$SU(4)=(U(1) \text{ X } F4) \text{ } (E6 \text{ X } SU(2)) \tag{8}$$

Origin of the supra-force:

The electromagnetism U(1) X F4 is attracted by the strong interactions G2 X SU(3) to give rise to the supra-force SU(5):

$$SU(5)=(U(1) \ X \ F4) \ (G2 \ X \ SU(3)) \tag{9}$$

Origin of the quantum fluctuations of the vacuum:

Electromagnetism U(1) X F4, weak interactions E6 X SU(2) and strong interactions G2 X SU(3) attract each other to give rise to quantum fluctuations of the vacuum SU(6):

$$SU(6)=(U(1) \ X \ F4) \ (E6 \ X \ SU(2)) \ (G2 \ X \ SU(3))=[(e^2 \\ X \ -e^2) \ (e^2 \ X \ -e^2)] \ X \ (-D \ X \ D)=[(+1 \ X \ -1) \ (+1 \ X \ -1)] \\ X \ (-1 \ X \ +1)=[- \ 1 \ X \ -1] \ X \ -1 \ \ X \text{ the constant } -1 =+1 \ X \\ +1=e^2 \ X \ e= e^3 \tag{10}$$

According to equation (10), 3 virtual pairs of particles and antiparticles (specifically the components of the 3 internal symmetries) are created and destroyed continually in vacuum to enable the functioning of the fundamental interactions. Likewise, the result obtained, the e3 charge, will allow us to obtain a particle called infinitely dense mass, which, according to the fundamental theory of charge, represents the fifth state or fifth phase of matter and plays a key role in the precise instant of the Big Bang.

Equations (5-10) explain the origin of 6 fundamental entities: electromagnetism, weak interactions, strong interactions, quantum gravity, supra-force and quantum fluctuations of the vacuum.

CHAPTER TWO
The behavior of the basic components of matter

With mathematical consistency and observing the conservation laws, it will be explained how the 6 fundamental leptons, which according to equation (10) are the components of quantum fluctuations of the vacuum, are combined by means of charge symmetry to first create each other, then generate the quarks and finally give rise to the messenger particles.

Behavior of the 6 fundamental leptons

Neutrino = $\dfrac{\text{Antineutrino -e X negative electrical charge of category 2 -}e^2}{\text{Positive electrical charge of category 2 }e^2}$

$$= \frac{-e\ X\ -e^2}{e^2} = e \tag{11}$$

Spin +1/2 of the neutrino = Spin +1/2 of the antineutrino (12)

Antineutrino= $\dfrac{\text{Neutrino e X positive electrical charge of category 2 }e^2}{\text{Negative electrical charge of category 2 -}e^2}$

$$= \frac{e\ X\ e^2}{-e^2} = -e \tag{13}$$

Spin +1/2 of the antineutrino=Spin +1/2 of the neutrino (14)

Levorotatory electron= $\dfrac{\text{neutrino -D X negative electrical charge of category 2 -}e^2}{\text{Positive magnetic charge D}}$

$$= \frac{-D \text{ X } -e^2}{D}$$

$$= \frac{-\text{Đ} \text{ X } -e^2}{\text{Đ}} = -e^2 \qquad (15)$$

Spin $+1/2$ of the levorotatory electron = Spin $+ 1/2$ of the $\qquad$ (16) neutrino

$$\text{Levorotatory positron} = \frac{\text{Antineutrino D X positive electrical charge of category 2 } e^2}{\text{Negative magnetic charge D}}$$

$$= \frac{D \text{ X } e^2}{-D}$$

$$= \frac{\text{Đ} \text{ X } e^2}{-\text{Đ}} = e^2 \qquad (17)$$

Spin$+1/2$ of the levorotatory positron=Spin $+1/2$ of the $\qquad$ (18) antineutrino

$$\text{Dextrorotatory electron} = \frac{\text{Dextrorotatory positron e2 X positive electrical charge of category 2 } e^2}{\text{Negative electrical charge of category 2 } -e^2}$$

$$\frac{=e^2 \text{ X } e^2}{-e^2} = -e^2 \qquad (19)$$

Spin $-1/2$ of the dextrorotatory electron = Spin $-1/2$ of the $\qquad$ (20) dextrorotatory positron

$$\text{Dextrorotatory positron} = \frac{\text{Dextrorotatory electron } -e^2 \text{ X negative electrical charge of category 2 } -e^2}{\text{Positive electrical charge of category 2 } e^2}$$

$$\frac{= -e^2 \text{ X } -e^2}{e^2} = e^2 \tag{21}$$

Spin -1/2 of the dextrorotatory positron = Spin -1/2 of the (22)
dextrorotatory electron

The 6 fundamental leptons and their quantum numbers	Magnetic charge	Electrical charge	Spin
Neutrino	-D (-1)	e	+1/2
Antineutrino	D (+1)	-e	+1/2
Levorotatory Electron	0	$-e^2$ (-1)	+1/2
Levorotatory Positron	0	e^2 (+1)	+1/2
Dextrorotatory Electron	0	$-e^2$ (-1)	-1/2
Dextrorotatory Positron	0	e^2 (+1)	-1/2

The behavior of quarks

The 3 heavy flavors t c b

$$t = \frac{\text{Dextrorotatory electron } -e^2 \text{ X levorotatory electron } -e^2}{\text{Dextrorotatory positron } e^2}$$

$$\frac{= -e^2 \text{ X } -e^2}{e^2} = e^2 \text{ (+1)} \tag{23}$$

Spin -1/2 of t=Spin -1/2 of the dextrorotatory electron + (24)
spin +1/2 of the levorotatory electron + spin -1/2 of the
dextrorotatory positron

$$c = \frac{\text{Dextrorotatory positron } e^2 \text{ X antineutrino D}}{\text{neutrino } -D}$$

$$\frac{= e^2 \text{ X D}}{-D} = \frac{e^2 \text{ X } \cancel{D}}{-\cancel{D}} = e^2 \text{ (+1)} \tag{25}$$

Spin +1/2 of c=Spin -1/2 of the dextrorotatory positron (26)
+1/2 of the Antineutrino + spin + 1/2 of the neutrino

$$b = \frac{\text{Dextrorotatory electron } -e^2 \text{ X neutrino } -D}{\text{Antineutrino D}}$$

$$= \frac{-e^2 \text{ X } -D}{D} = \frac{-e^2 \text{ X } -Đ}{Đ} = e^2 \, (-1) \tag{27}$$

Spin + 1/2 of b=Spin -1/2 of dextrorotatory electron + (28)
spin +1/2 of the neutrino + spin + 1/2 of the antineutrino

The 3 light flavors u d s

$$u = \frac{\text{Neutrino-D X dextrorotatory electron } -e^2}{\text{Dextrorotatory positron } e^2}$$

$$= \frac{-D \text{ X } -e^2}{e^2} = \frac{-D \text{ X } - e^2}{e^2} = -D \, (-1) \tag{29}$$

Spin -1/2 of u= Spin +1/2 of the neutrino + spin -1/2 of the (30)
dextrorotatory electron + spin -1/2 of the dextrorotatory
positron

$$d = \frac{\text{Antineutrino D X dextrorotatory positron } e^2}{\text{Levorotatory electron } -e^2}$$

$$= \frac{D \text{ X } e^2}{-e^2} = \frac{D \text{ X } e^2}{-e^2} = D \, (+1) \tag{31}$$

Spin +1/2 of d=Spin +1/2 of the antineutrino + spin -1/2 of (32)
the dextrorotatory positron + spin +1/2 of the levorotatory
electron

$$s = \frac{\text{Antineutrino D X levorotatory positron } e^2}{\text{Dextrorotatory electron } -e^2}$$

$$= \frac{D \text{ X } e^2}{-e^2} = \frac{D \text{ X } e^2}{-e^2} = D \, (+1) \tag{33}$$

Spin +1/2 of s=Spin+1/2 of the antineutrino + spin+1/2 of (34)
the levorotatory positron + spin -1/2 of the dextrorotatory
electron

Based on equations (23), (25), (27), (29), (31) and (33), it
is possible to calculate with mathematical consistency, not
only the fractional electrical charges of the 6 flavors, but also
their fractional magnetic charges. When calculating the fractional electrical charge of quarks, the magnetic charge factor must be considered. Meanwhile, when determining their
fractional magnetic charges, the electric charge factor must
also be taken into account.

Calculation of the fractional electrical charges of the 6 flavors

The 3 heavy flavors t c b

Electrical charge of t=+1(equation (23))

Electrical charge of c=+1(equation (25))

Electrical charge from b=-1(equation (27))

According to equations (23), (25) and (27), t c b has a total
of 0 magnetic charge. Therefore, their respective electrical
charges are added together with fractions totaling 0: - 1/3 +
(-1/3) + 2/3.

Fractional electric charge of t=+1 + (-1/3) =2/3   (35)

Fractional electric charge of c=+1 +(-1/3) =2/3   (36)

Fractional electric charge of b= -1 + 2/3= -1/3   (37)

The 3 light flavors u d s

Magnetic charge of u=-1(equation (29))

Magnetic charge of d=+1(equation (31))

Magnetic charge of s=+1(equation (33))

According to equations (29), (31) and (33), the magnetic charges of u d s are respectively -1, +1 and +1, which, when multiplied, result in -1. Therefore, these charges are added together with fractions totaling -1: 1/3 + (-2/3) + (-2/3) and, since they are magnetic charges which are converted into fractional electrical charges, each result obtained is multiplied by the constant -1.

Fractional electric charge of u=-1 + 1/3=-2/3 X the
constant - 1=2/3   (38)

Fractional electric charge of d=+1 +(-2/3) =1/3 X the
constant -1=-1/3   (39)

Fractional electric charge of s=+1 +(-2/3) =1/3 X the
constant -1=-1/3   (40)

Calculation of the fractional magnetic charges of the 6 flavors

The 3 heavy flavors t c b

According to equations (23), (25) and (27), the electrical charges of t c b are respectively +1, +1 and -1, which, when multiplied, result in -1. Therefore, these charges are added together with fractions totaling -1: -2/3 + (-2/3) + 1/3 and, since they are electrical charges which are converted into fractional magnetic charges, each result obtained is multiplied by the constant -1.

Fractional magnetic charge of t=+1 +(-2/3) =1/3 X the
constant -1=-1/3 (41)

Fractional magnetic charge of c=+1 + (-2/3) =1/3 X the
constant -1=-1/3 (42)

Fractional magnetic charge of b=-1 + 1/3=-2/3 X the
constant -1=2/3 (43)

The 3 light flavors u d s

According to equations (29), (31) and (33), u d s has a total
electrical charge of 0. Therefore, their respective magnetic
charges are added together with fractions totaling 0 (2/3)
+(-1/3) + (-1/3).

Fractional magnetic charge of u=-1 + 2/3=-1/3 (44)

Fractional magnetic charge of d=+1 +(-1/3) =2/3 (45)

Fractional magnetic charge of s=+1 + (-1/3) =2/3 (46)

The 6 flavors and their quantum numbers	Fractional magnetic charge	Fractional electric charge	Spin
t	-1/3	2/3	-1/2
c	-1/3	2/3	+1/2
b	2/3	-1/3	+1/2
u	-1/3	2/3	-1/2
d	2/3	-1/3	+1/2
s	2/3	-1/3	+1/2

Thanks to the calculations that had just been made, it is possible to explain, with mathematical consistency and observance of the laws of conservation, the behavior of the two
fundamental baryons, the proton and the neutron.

Proton behavior

3 components: uud

Electrical charge of the proton=2/3+ 2/3 + (-1/3) =3/3=+1=e^2 (47)

The proton›s magnetic charge=-1/3+ (-1/3) + 2/3=0   (48)

Proton Spin -1/2 = Spin -1/2 of u + spin -1/2 of u + spin   (49)
+1/2 of d

Neutron behavior

3 components: udd

Electrical charge of the neutron=2/3 + (-1/3) + (-1/3)=0   (50)

The neutron's magnetic charge=-1/3 + 2/3 + 2/3=3/3=+1 X (51)
the constant -1=-1=-e

Note that while equation (50) suggests that the neutron has 0 electrical charge of category 2, formula (51) proposes a negative electrical charge of category 1, which, as we will see later, has serious consequences at both theoretical and experimental levels.

Spin +1/2 of the neutron = Spin -1/2 of u +spin +1/2 of d + (52)
spin +1/2 of d

The 2 fundamental baryons and their quantum numbers	Magnetic charge	Electrical charge	Spin
Proton	0	e^2 (+1)	-1/2
Neutron	D (+1)	-e	+1/2

Based on the previous approaches, it will be possible in the following chapter to coherently explain 2 great phenomena within the atomic nucleus: the weak interactions, which favor the conversion of the neutron into a proton and vice versa, as well as the strong interactions, which maintain the cohesion of the nucleus.

Behavior of the messenger particles

According to the fundamental theory of charge, electromagnetism, weak interactions, strong interactions, quantum gravity and supra-force work by a total of 9 messenger particles, which are generated from the phenomenon called quantum vacuum fluctuations where 3 virtual pairs of particles and antiparticles (the 6 fundamental leptons) are created and destroyed continually.

9 messenger particles:

Virtual photon (electromagnetism)

Particles w-, w+ and z, Higgs boson and Goldstone boson (weak interactions)

Gluon (strong interactions)

Graviton (quantum gravity)

Inflaton (supra-force)

Virtual photon behavior

Virtual photon = $\dfrac{\text{Levorotatory electron } -e^2 \text{ X levorotatory positron } e^2}{\text{Negative magnetic charge } -D \text{ X Positive magnetic charge } D}$

$$= \frac{-e^2 \text{ X } e^2}{-D \text{ X } D} = \frac{-1 \text{ X } +1}{-1 \text{ X } +1} = \frac{-1}{-1 \text{ X la constante } -1} = \frac{-1}{+1}$$

$$= \frac{-e^2}{e} = -e \tag{53}$$

Spin +1 of the virtual photon=Spin +1/2 of the levorotatory (54) electron + spin + 1/2 of the levorotatory positron

Behavior of the particles w⁻, w⁺ and z, the Higgs boson and the Goldstone boson

Particle W⁻= $\dfrac{\text{Levorotatory electron } -e2 \text{ X Positive electrical charge of category 2 } e^2}{\text{Levorotatory positron } e^2}$

$$= \frac{-e^2 \text{ X } e^2}{e^2} = -e^2 \tag{55}$$

Spin +1 of w-=Spin +1/2 of the levorotatory electron + (56) spin+1/2 of the levorotatory positron

Particle w⁺= $\dfrac{\text{Dextrorotatory positron } e^2 \text{ X negative electrical charge of category 2 } -e^2}{\text{Dextrorotatory electron } -e^2}$

$$= \frac{-e^2 \text{ X } -e^2}{-e^2} = e^2 \tag{57}$$

Spin -1 of w+=Spin -1/2 of the dextrorotatory positron + (58)
spin -1/2 of the dextrorotatory electron

$$\text{Particle } z = \frac{\text{Levorotatory Electron } -e^2 \text{ X negative magnetic charge } -D}{\text{Dextrorotatory positron } e^2}$$

$$= \frac{-e^2 \text{ X } -D}{e^2} = \frac{-e^2 \text{ X } -D}{e^2} = -D = -1 \text{ X the constant-}1 = +1 = e \quad (59)$$

Spin 0 of z=Spin +1/2 of the levorotatory electron + spin (60)
-1/2 of the dextrorotatory positron

$$\text{Higgs boson HB} = -w^+ e^2 \text{ X } z e \frac{}{w^- -e^2} = \frac{e^2 \text{ X } e}{-e^2} = -e \quad (61)$$

Spin 0 of HB=Spin -1 of w$^+$ + spin 0 of z + spin +1 of w$^-$ (62)

According to the fundamental theory of charge, at high energy, for example inside a black hole, weak interactions, through the Goldstone boson, cause the decay of the three atomic components, the electron, the proton and the neutron, generating an infinitely dense mass that does not allow visible light to escape from the event horizon. This phenomenon will be explained in chapter six.

$$\text{Goldstone boson GB} = \frac{w^- -e^2 \text{ X } w^+ e^2}{\text{HB } -e \text{ X } z e} = \frac{-e^2 \text{ X } e^2}{-e \text{ X } e} = e^2 \quad (63)$$

Spin 0 from GB=Spin +1 from w- + spin -1 from w$^+$ + spin (64)
0 from HB + spin 0 from z

Gluon behavior

$$\text{Gluon} = \frac{\text{Levorotatory electron } -e^2 \text{ X positive electrical charge of Category 2 } e^2}{\text{Antineutrino } -e}$$

$$= \frac{-e^2 \text{ X } e^2}{-e} = e^3 \qquad (65)$$

It should be noted that the positive electrical charge of a gluon, category 3, enables it to fulfil its function as a messenger particle of strong interactions, as will be seen in the following chapter.

Gluon Spin +1= Spin +1/2 of the levorotatory electron + spin +1/2 of the antineutrino (66)

Graviton behavior

$$\text{Graviton} = \frac{\text{Electron levorotatory } -e^2 \text{ X positron levorotatory } e^2}{\text{Neutrino } -D \text{ X antineutrino } D}$$

$$= \frac{-e^2 \text{ X } e^2}{-D \text{ X } D} = \frac{-1 \text{ X } +1}{-1 \text{ X } +1} = \frac{-1}{-1 \text{ X la constante } -1} = \frac{-1}{+1}$$

$$= \frac{-e^2}{e} = -e \qquad (67)$$

spin +2 of graviton = Spin +1/2 of the levorotatory electron (68) + spin+1/2 of the levorotatory positron + spin+1/2 of the neutrino + spin+1/2 of the antineutrino

Inflaton behavior

At relatively high energy, for example, within the solar nucleus, the fifth fundamental interaction of contemporary physics occurs, the supra- force, which, through its messenger particle, the inflaton, converts positive masses into positive energy in the form of ultraviolet light.

Inflaton= (Dextrorotatory electron $-e^2$ X levorotatory positron e^2 X dextrorotatory positron e^2) (Antineutrino D)

$$= (-e^2 \text{ X } e^2 \text{ X } e^2)\,(D)= (-1 \text{ X } +1 \text{ X } +1)\,(+1) = (-1) \qquad (69)$$
$$(+1 \text{ X la constante } -1) = (-1)\,(-1) = (-e^2)\,(-e) = e^3$$

Inflaton Spin 0 = Spin $-1/2$ of the dextrorotatory electron (70)
+ spin $+1/2$ of the levorotatory positron + spin $-1/2$ of the dextrorotatory positron + spin $-1/2$ of the antineutrino

Equations (11-70), which describe the behavior of the basic components of matter, have mathematical consistency, as well observe the laws of conservation of charge and angular momentum, which, according to the fundamental theory of charges, constitute the only exact symmetries of Nature, since they are not broken at high or low energies. But these equations also reflect a strict order that is as follows:

1. When an even number of particles is involved in the formula with the presence of an even number of charges, the entity described has a mass of 0 (equations (53), (63) and (67)).

2. When an odd number of particles is involved in the formula, the entity described acquires mass (equations (11), (13), (15), (17), (19), (21), (23), (25), (27), (29), (31), (33) and (61)).

3. When an even number of particles is involved in the formula in the presence of an odd number of charges, the entity described acquires mass (equations (55), (57) and (59)).

4. When the entity described has an electrical charge of more than category 2, its mass is 0 regardless of its composition (equations (65) and (69)).

The 9 messenger particles and their quantum numbers	Fundamental interaction	Magnetic charge	Electrical charge	Spin
Virtual photon	Electromagnetism	D (+1)	-e	+1
w$^+$	Weak interactions	0	e^2 (+1)	-1
w$^-$	Weak interactions	0	-e^2 (-1)	+1
z	Weak interactions	-D (-1)	e	0
Higgs boson	Weak interactions	D (+1)	-e	0
Goldstone Boson	Weak interactions	0	e^2 (+1)	0
Gluon	Strong interactions	0	e^3	+1
Graviton	Quantum Gravity	D (+1)	-e	+2
Inflaton	Supra-force	0	e^3	0

CHAPTER THREE
The 3 internal symmetries

Description of electromagnetism

According to equation (53), the messenger particle of electromagnetism, the virtual photon, has a negative electrical charge of category 1, which allows it to fulfill its mediating function.

Interaction between the levorotatory electron $-e^2$ and the proton e^2 by means of the virtual photon -e, generating an attractive charge e:

$$e= \frac{-e^2 \, X -e}{e^2} \tag{71}$$

According to equation (71), the virtual photon, more than a mediating field, constitutes a mediating charge that favors the attraction between the 2 fermions.

Interaction between the levorotatory electron $-e^2$ and the proton e^2 by means of the virtual photon -e, generating an attractive charge e:

$$-e= \frac{-e^2 \, X -e}{-e^2} \tag{72}$$

Interaction between the levorotatory electron $-e^2$ and the neutrino e by means of the virtual photon -e, generating an attractive charge e^2:

$$e^2= \frac{-e^2 \, X -e}{e} \tag{73}$$

Equation (73) explains why, in the experiments, the electron is usually accompanied by the neutrino, producing the phenomenon known as the electronic neutrino.

Interaction between the levorotatory electron -e² and the neutron -e by means of the virtual photon -e, generating a repulsive charge -e²:

$$-e^2 = \frac{-e^2 \, X \, -e}{-e} \qquad (74)$$

Equation (74) is predicting an experimentally verifiable phenomenon: the repulsion between an electron and a neutron due to their negative electrical charges.

Description of weak interactions

According to the fundamental theory of charges, at low energies and at relatively high energies, weak interactions work through 4 massive messenger particles and, therefore, of short range: the particles w^+, w^- and z, as well as the Higgs boson. The purpose is to prevent the decay of the particles from violating the 2 exact symmetries of Nature, the laws of conservation of charge and angular momentum, as can be seen below.

Negative beta disintegration or neutron decay

According to the fundamental theory of charge, the neutron, possessing a positive magnetic charge D (+1), when decaying, decomposes, not to the particle w-, but to the Higgs boson, which also has a positive magnetic charge D (+1), disintegrating into antineutrino D (+1), levorotatory electron -e² and proton e²:

$$\text{Decayed neutron} = \frac{\text{Higgs boson D X antineutrino D X electron Levorotatory } -e^2}{\text{Proton } e^2}$$

$$= \frac{D \, X \, D \, X \, -e^2}{e^2} = \frac{D \, X \, D \, X \, -e^2}{e^2}$$

$$= D \, X \, D = +1 \, X \, +1 = +1 \, X \text{ the constant } -1 = -1 = -e \qquad (75)$$

Spin +1/2 of the decayed neutron =Spin 0 of the Higgs $\quad$ (76)
boson + spin +1/2 of the antineutrino + spin +1/2 of the
levorotatory electron + spin -1/2 of the proton

According to equations (75) and (76), the neutron, when using the Higgs boson to decay into an antineutrino, a levorotatory electron and a proton, observes the laws of conservation of charge and angular momentum, the only two exact
symmetries of Nature.

Positive beta disintegration or decay of the nuclear proton

The nuclear proton, which has a positive electrical charge of
category 2 $e^2(+1)$, uses the particle w+ $e^2(+1)$ to decay to a
dextrorotatory positron $e^2(+1)$, neutrino -D and neutron D:

Decayed nuclear proton $= \dfrac{w^+ \; e^2 \; X \; \text{dextrorotatory positron } e^2 \; X \text{ neutrino } -D}{\text{Neutron D}}$

$$= \frac{e^2 \; X \; e^2 \; X \; \text{-D}}{D} = \frac{e^2 \; X \; e^2 \; X \; \text{-Đ}}{Đ}$$

$$= e^2 \; X \; e^2 = +1 \; X \; +1 = +1 = e^2 \qquad (77)$$

Spin -1/2 of the decayed nuclear proton =Spin -1 of w^+ + $\quad$ (78)
spin -1/2 of the dextrorotatory positron + spin +1/2 of the
neutrino + spin + 1/2 of the neutron

When a proton and a neutron within the atomic nucleus decay, the quark u becomes the quark d and vice versa. According to the fundamental theory of charge, a total of 4
messenger particles are involved in these 2 processes: w^+,
w^-, z and the Higgs boson (HB), ensuring that the laws of
electrical charge, magnetic charge and angular momentum
are observed.

Conversion of u to d via w⁺ and z

Law on the conservation of electrical charge

$$u(2/3) = w^+(+1) + z(0) + d(-1/3) \qquad (79)$$

Magnetic charge preservation law

$$u(-1/3) = w^+(0) + z(-1) + d(2/3) \qquad (80)$$

Law of conservation of angular momentum

$$u(\text{spin } -1/2) = w^+(\text{spin } -1) + z(\text{spin } 0) + d(\text{spin } +1/2) \qquad (81)$$

Conversion of d to u via w⁻ and HB

Law on the conservation of electrical charge

$$d(-1/3) = w^-(-1) + HB(0) + u(2/3) \qquad (82)$$

Magnetic charge preservation law

$$d(2/3) = w^-(0) + HB(+1) + u(-1/3) \qquad (83)$$

Law of conservation of angular momentum

$$d(\text{spin } +1/2) = w^-(\text{spin } +1) + HB(\text{spin } 0) + u(\text{spin } -1/2) \qquad (84)$$

Description of the strong interactions

According to the fundamental theory of charge, within the atomic nucleus, strong interactions function by means of 2 factors:

1.Gluon, which according to equation (65) has a positive electrical charge of category 3 e^3, promotes the attraction between protons e^2 and neutrons $-e$.

2.The opposite spins -1/2 and +1/2 of the protons and neutrons, which when multiplied by the value 2, result in -1, which indicates negative electric charge of category 1 -e, and +1, which indicates positive electric charge of category

1 e. These opposite charges cause protons and neutrons to be additionally attracted by the messenger particle of electromagnetism, the virtual photon.

Calculations of strong interactions within the atomic nucleus

First phase
The protons e^2 and neutrons -e attract each other by gluons e^3:

$$\frac{e^2 \times e^3}{-e} = -e^4 \qquad (85)$$

Second phase
The protons e^2 repel each other through virtual photons -e:

$$\frac{e^2 \times -e}{e^2} = -e \qquad (86)$$

Third phase
Neutrons -e reject each other by means of virtual photons -e:

$$\frac{-e \times -e}{-e} = -e \qquad (87)$$

Fourth phase
The results of equations (86) and (87) are multiplied:

$$-e \times -e = e^2 \qquad (88)$$

Fifth phase
The result of equation (85) is divided by the result of equation (88):

$$\frac{-e^4}{e^2} = -e^2 \qquad (89)$$

Sixth phase
The protons and neutrons, due to their respective spins -1/2 and + 1/2, are converted into opposite electrical charges of category 1 -e and e, are additionally attracted by virtual photons -e:

$$\frac{e \, X -e}{-e} = e \tag{90}$$

Seventh phase
The result of equation (89) is multiplied by the result of equation (90):

$$-e^2 \, X \, e = -e^3 \tag{91}$$

The result of equation (91) implies that the atomic nucleus is held together by a negative category 3 electrical charge.

Calculations of strong interactions within the proton (uud)

First phase
The product of the uu -1/2 spins converted into -e and -e is divided by the +1/2 spin of d converted into e:

$$\frac{-e \, X -e}{e} = e \tag{92}$$

Second phase
The result of equation (92) is multiplied by the charge e^3 of the gluons:

$$e \, X \, e^3 = e^4 \tag{93}$$

The result of equation (93) implies that the proton is held together by a positive electrical charge of category 4.

Calculations of strong interactions within the neutron (udd)

First phase
The product of the dd +1/2 spins converted to e and e is divided by the -1/2 u spin converted to -e:

$$\frac{e \, X \, e}{-e} = -e \tag{94}$$

Second phase
The result of equation (94) is multiplied by the charge e^3 of the gluons

$$-e \, X \, e^3 = -e^4 \tag{95}$$

The result of equation (95) implies that the neutron is held together by a negative electrical charge of category 4.

It should be noted that when the particle is held together by a negative electrical charge (equations (91) and (95)), it is unstable, but if it is held together by a positive electrical charge (equation (93)) it is stable.

The confinement of quarks and gluons

Based on the calculations of strong interactions that have been made, it is possible to explain coherently why quarks and gluons are confined.

Confinement within the atomic nucleus

The negative electrical charge of category 3, which according to equation (91) keeps the atomic nucleus cohesive, prevents the gluons and protons, whose electrical charges are positive, from escaping, confining them within the nucleus, yet favors the neutrons, whose electrical charge is negative, to eventually leave freely.

Confinement within the proton uud

The positive electric charge of category 4, which according to equation (93) keeps the proton cohesive, prevents uu, whose -1/2 spins become a negative electric charges of category 1, from escaping. In turn, uu retains both d, whose spin +1/2 becomes a positive electrical charge of category 1, and gluons.

Confinement within the neutron udd

The negative electrical charge of category 4, which according to equation (95) keeps the neutron cohesive, prevents dd and gluons from escaping. In turn, they retain u.

CHAPTER FOUR
The 2 external symmetries

Description of Quantum Gravity

According to the fundamental theory of charge, quantum gravity is the attraction between 2 positive masses through a negative energy quantized by a graviton, generating, more than a gravitational field, a negative gravitational energy.

Positive mass

The atom, the basic component of every object, has, according to the following equation, a positive electrical charge of category 1. Therefore, every object, from a star to a grain of sand, has a positive mass.

$$\text{Atom} = \frac{\dfrac{\text{Levorotatory electron } -e^2}{\text{Proton } e^2}}{\text{Neutron } -e}$$

$$= \frac{\dfrac{-e^2}{e^2}}{-e}$$

$$= \frac{-e^2 \times -e}{e^2} = e \qquad\qquad (96)$$

In order to calculate the spin of the atom on the basis of equation (96), we must bear in mind a phenomenon called charge-particle duality, by virtue of which the proton, at the nuclear level, is a spin $-1/2$ particle, but, at the atomic level, it behaves like a simple positive electric charge of category 2, devoid of spin.

Spin +1 of atom=Spin +1/2 of levorotatory electron + spin   (97)
0 of atomic proton + spin + 1/2 of neutron

It is important to underline that the spin +1 of the atom is inherited by every object made up of atoms, which, as we will see later on, will allow us to explain coherently why spacetime has 4 dimensions.

Negative energy

According to equation (67), graviton, generated from the quantum fluctuations of vacuum by 2 virtual pairs of particles and antiparticles, whose spins add up to +2, has a negative electrical charge of category 1, which, according to equation (1), is equivalent to a magnetic charge, which implies that the energy it holds, quantized, is negative and equal to the mass of the magnetic monopole, which is approximately 10^{14} Gev.

The distance factor

The distance between 2 objects attracted by quantum gravity is not squared due to the presence of the graviton.

Negative gravitational energy

The gravitational attraction between 2 positive masses through a negative energy generates not a gravitational field, but a negative gravitational energy.

The Law of Quantum Gravity

Based on the above, the law of quantum gravity is enunciated: the negative gravitational energy -EG that attracts 2 objects is directly proportional to their positive masses M_1, M_2 and inversely proportional to both the negative energy -e of the gravitons that mediate between these objects, and to the distance R which separates them.

$$-E_G = \frac{M_1 \times M_2}{-e \times R} \tag{98}$$

Spin +1 of M1 + spin +1 of M2 + spin +2 of -e(graviton)   (99)
=Spin +4

The spin +4 of equation (99) determines that spacetime, which, together with gravitation, are two different manifestations of the same phenomenon, has precisely four dimensions.

Equation (98) can be used to calculate the gravitational effects at small and large scales, from the fall of an apple to the ground to the attraction between the Sun and each of its planets to the precession in the perihelion of Mercury or the interaction between two galaxies.

Description of the supra-force

According to equation (9), electromagnetism, when unified with strong interactions, generates supra-force, whose messenger particle, the inflaton, has, according to equation (69), a positive electrical charge of category 3 e^3, which, within the solar core for example, violently repels positive masses e in order to produce positive energy e^4 in the form of ultraviolet light:

$$e^4 = e \times e^3 \tag{100}$$

The phenomenon described by equation (100) shows that in this low energy Universe the supra-force is still fully valid.

Photoelectric effect

The phenomenon called the photoelectric effect can be explained based on the fundamental theory of charge:

The ultraviolet light e^4 attracts, through virtual photons -e, the electrons $-e^2$ of a charged metal, converting it into an object with a positive electrical charge of category 1 e.

$$e = \frac{e^4}{-e^2 \, X \, -e} \tag{101}$$

The infinitely dense mass

As we know, there are 5 exceptional Lie groups: G2, F4, E6, E7 and E8. In the first chapter, we saw that the first 3 represent the neutrino, the dextrorotatory electron and the dextrorotatory positron respectively. So, according to the fundamental theory of charges, the other 2 are electrical charges of category 2: E7, negative and E8, positive.

The quantum fluctuations of vacuum SU(6), which, according to equation (10), have a positive electrical charge of category 3 e^3, are always framed by an infinitely dense point, are also unified with a negative electrical charge of category 2 E7 $-e^2$ and generate the infinitely dense mass SU(8):

$$SU(8) = \frac{SU(6)}{E7} = \frac{e^3}{-e^2} = -e \tag{102}$$

Within an object of high energies, such as a black hole, for example, the infinitely dense mass -e is violently attracted by the inflaton e3 to create negative energy -e4 in the form of X-rays:

$$-e^4 = -e \, X \, e^3 \tag{103}$$

While ultraviolet light, due to its positive electrical charge of category 4, is able to damage the skin and, similar to an atom, has a positive electrical charge of category 1, X-rays, with a negative electrical charge of category 4, are absorbed by it to generate radiographic plates. This constitutes solid experimental support in favor of these various approaches.

So far, 5 fundamental interactions have been described: electromagnetism, weak interactions, strong interactions, quantum gravity and supra-force, operating through messenger particles whose electrical charges are of category 1, 2 or 3. In the next chapter, 2 other fundamental interactions will be explained: the speed of light and the cosmological constant, which use, instead of messenger particles, messenger charges to generate gamma rays and an expanding space-time, respectively.

CHAPTER FIVE
The grand event

Throughout this piece, within an infinitely dense point and therefore at high energies, some of the following components of the cosmic egg have been generated, from which the grand event called the Big Bang was produced: quantum gravity SU(4), supra-force SU(5), quantum fluctuations of vacuum SU(6) and infinitely dense mass SU(8). Three other fundamental entities are missing to complete the picture: 10-dimensional space SU(10), 11-dimensional spacetime SU(11) and the constant speed of light SU(12).

10 dimensional space SU(10)

The unification between infinitely dense mass SU(8) -e and a positive electrical charge of category 2 E8 e^2 produces 10-dimensional space SU(10):

$$SU(10)=SU(8) \times E8=-e \times e^2=-e^3 \qquad (104)$$

11-dimensional spacetime SU(11)

The interaction between 10-dimensional space SU(10) $-e^3$ and a positive magnetic charge SU(3) or a negative electrical charge of category 1 -e gives rise to 11-dimensional spacetime SU(11):

$$SU(11)=SU(10) \times SU(3)=-e^3 \times -e=e^4 \qquad (105)$$

The constant speed of light SU(12)

The interaction between 11-dimensional spacetime SU(11) e^4 and a negative magnetic charge G2 or a positive electrical charge of category 1 e generates the constant speed of light SU(12):

$$SU(12) = \frac{SU(11)}{G2} = \frac{e^4}{e} = e^3 \tag{106}$$

Within the solar nucleus, for example, the constant speed of light, as the sixth fundamental interaction of contemporary physics, uses its positive electric charge of category 3 e^3 as a mediating factor to violently repel positive masses e, generating positive energy e^4 in the form of gamma rays:

$$e^4 = e^3 \text{ X e} \tag{107}$$

The 7 protagonists of the Big Bang and their quantum numbers	Electrical charge	Spin	Associated Lie symmetry
Graviton (quantum gravity)	-e	+2	SU(4)
Inflaton (supra-force)	e^3	0	SU(5)
Quantum fluctuations of vacuum	e^3	+1	SU(6)
Infinitely dense mass	-e	+1	SU(8)
10 dimensional space	$-e^3$	0	SU(10)
11 dimensional spacetime	e^4	0	SU(11)
Constant speed of light	e^3	0	SU(12)

The Big Bang

By having the 7 components of the cosmic egg, it is possible to explain coherently the details of the Big Bang that gave rise to the current Universe. This great event consists of 7 stages:

First stage
Within 11-dimensional spacetime SU(11), the infinitely dense mass SU(8) -e is violently attracted by the inflaton (supra-force) SU(5) e^3, generating primary X-rays SU(13):

$$SU(13)=SU(8) \text{ X } SU(5)= -e \text{ X } e^3=-e^4 \tag{108}$$

Spin +1 of the primary X rays = Spin +1 of the infinitely dense mass + spin 0 of the inflaton $\tag{109}$

These primary X rays, when traveling at speeds greater than light in the framework of 11-dimensional spacetime, provide an inflationary Universe.

Second stage
When infinitely dense mass disintegrates, its contents, the 6 fundamental leptons, come into play: both the dextrorotatory electron- levorotatory positron pair and the dextrorotatory positron- levorotatory electron pair to become pure energy, due to their opposite charges and spins. However, this is not the case with the remaining antineutrino-neutrino pair because, although both have opposite charges, their spins are equal (+1/2). On the contrary, they become a mass with a positive electrical charge of category 1, according to the following equation:

Antineutrino SU(3)
    X neutrino G2=Positive magnetic charge D X
        negative magnetic charge -D
        =D X -D=+1 X -1=-1 X lthe constant
        -1=+1=e $\tag{110}$

Spin +1/2 of the antineutrino + spin +1/2 of the neutrino = (111)
Spin +1 of the positive mass

Within the framework of 10-dimensional space SU(10), the positive mass e, consisting of antineutrinos SU(3) and neutrinos G2, is violently repelled by the constant speed of light SU(12) e^3, producing primary gamma rays Su(14):

$$SU(14)=SU(12) \times (SU(3) \times G2)=e^3 \times e=e^4 \tag{112}$$

Spin +1 of the primary gamma rays =Spin 0 of the (113)
constant speed of light + spin +1 of the positive mass

These primary gamma rays, traveling at speeds greater than the speed of light, within the framework of 10-dimensional space, feed back into the inflationary Universe.

Third stage
The primary X rays SU(13) -e^4 interact with the inflaton SU(5) e^3, giving rise to the dark matter or primary antimatter SU(16):

$$SU(16) = \frac{SU(13)}{SU(5)} = \frac{-e^4}{e^3} = -e \tag{114}$$

Dark Matter Spin +1 = Primary X Ray Spin +1 + Inflaton (115)
Spin 0

Fourth stage
The primary gamma rays SU(14) e^4 interact with the constant speed of light SU(12) e^3, generating positive matter or observable matter SU(17):

$$SU(17) = \frac{SU(14)}{SU(12)} = \frac{e^4}{e^3} = e \tag{116}$$

Spin +1 of the positive matter = Spin +1 of the primary (117)
gamma rays + spin 0 of the constant speed of light

The appearance of dark matter and positive matter marks the end of the inflationary Universe.

Fifth stage
Dark matter SU(16) -e and positive matter SU(17) e attract each other by means of graviton SU(4) -e, to form 4-dimensional resting spacetime SU(18):

$$SU(18) = \frac{SU(16) \times SU(17)}{SU(4)} = \frac{-e \times e}{-e} = e \tag{118}$$

Note that, at a more fundamental level, the graviton, instead of a negative energy, uses a negative electrical charge of category 1 to perform its mediating function.

4 dimensions of spacetime = Spin +1 of the dark matter + spin +1 of the positive matter + spin +2 of the graviton (119)

4-dimensional spacetime comes to replace the 11-dimensional spacetime and 10-dimensional space of the early Universe, becoming the new framework for events.

Sixth stage
On one hand, the dark matter SU(16) -e is attracted by the inflaton SU(5) e^3 and on the other hand, the same thing happens between the positive matter SU(17) e and the graviton SU(4) -e, originating the cosmological constant SU(21):

$$SU(21) = \frac{SU(16) \times SU(5)}{SU(17) \times SU(4)} = \frac{-e \times e^3}{e \times -e} = e^2 \tag{120}$$

The cosmological constant represents the seventh fundamental interaction of the contemporary physics and, like the constant speed of light, is mediated, not by a messenger particle, but by a messenger charge.

Seventh stage

The cosmological constant SU(21) e^2 drives resting 4-dimensional spacetime SU(18) e, creating an expanding 4-dimensional spacetime SU(22):

$$SU(22)=SU(21) \times SU(18)=e^2 \times e=e^3 \qquad (121)$$

With the emergence of expanding 4-dimensional spacetime, the Big Bang is completed.

The 7 experimental consequences of the Big Bang and their quantum numbers	Electrical charge	Spin	Associated Lie symmetry
Primary X rays	$-e^4$	+1	SU(13)
Primary gamma rays	e^4	+1	SU(14)
Dark matter or primary antimatter	$-e$	+1	SU(16)
Positive matter or observable matter	e	+1	SU(17)
4-dimensional space time at rest	e	+4=4 dimensions	SU(18)
Cosmological constant	e^2	0	SU(21)
Expanding 4-dimensional time space	e^3	+4=4 dimensions	SU(22)

The electromagnetic spectrum

According to equations (100), (103) and (107), high-energy photons, ultraviolet light, x rays and gamma rays, have respectively e^4, $-e^4$ and e^4 charges. According to the fundamental theory of charge, the remaining components of the electromagnetic spectrum are generated from them.

$$\text{Visible light} = \frac{\text{Ultraviolet light } e^4}{\text{Inflaton } e^3} = \frac{e^4}{e^3} = e \qquad (122)$$

$$\text{Radio} = \frac{\text{Gamma rays } e^4}{C\ e^3} = \frac{e^4}{e^3} = e \qquad (123)$$

Where C is the constant speed of light

$$\text{Infrared light} = \frac{\text{X-ray } -e^4}{\text{Inflaton } e^3} = \frac{-e^4}{e^3} = -e \qquad (124)$$

According to equation (114), dark matter has a negative electrical charge of category 1. Consequently, unlike positive matter (galaxies, stars and planets), which reflects visible light and allow it to be observed visually, dark matter absorbs this light and it becomes invisible. However, thanks to infrared light, it is going to be possible to detect it. Because of its identical electrical charges, dark matter tends to reflect the light in question, which, when captured by special equipment, can reveal the presence of this elusive entity, which would represent solid experimental support for the fundamental theory of charge.

The 6 components of the electromagnetic spectrum and their quantum numbers	Electrical charge	Spin
Ultraviolet light	e^4	+1
X rays	$-e^4$	+1
Gamma rays	e^4	+1
Visible light	e	+1
Infrared light	$-e$	+1
Radio	e	+1

Description of the hydrogen atom

The hydrogen atom (the most abundant matter in the Universe) has as its nucleus a proton orbited by a levorotatory electron. According to equation (15), this is composed of a neutrino with a negative magnetic charge -D in the presence of a negative electrical charge of category 2 -e^2 and a positive magnetic charge D.

$$H = \frac{(-D \times D)(-e^2)}{e^2} = \frac{(-D \times D)(-e^2)}{e^2} = (-D \times D)$$
$$=-1 \times +1 = -1 \times \text{ the constant } -1 = +1 = e \qquad (125)$$

In equation (125), the negative electrical charge of category 2 -e^2 of the levorotatory electron, when neutralized with the opposite charge e^2 of the proton, allows the two opposite magnetic charges -D and D to stop neutralizing and multiply, resulting in a positive electrical charge of category 1 e. During the process, the positive magnetic charge D, by virtue of the charge-particle duality, becomes an antineutrino.

Spin +1 of H=Spin +1/2 of neutrino + spin +1/2 of   (126)
antineutrino + spin 0 of the atomic proton

As the reader, always remember that, based on the charge-particle duality, the proton, at the nuclear level, is a spin -1/2 particle, but, at the atomic level, it behaves as a simple charge devoid of spin.

The charge-particle duality, proposed by the fundamental theory of charge, has the following experimental support: in the laboratory it is usually observed that a neutrino suddenly becomes an antineutrino and vice versa. What occurs, is that the neutrino, due to its negative magnetic charge, is always accompanied by a positive magnetic charge, and, by virtue of the charge-particle duality, tends to become an antineutrino while the neutrino is transformed into a simple negative magnetic charge. Of course, that process is reversible.

CHAPTER SIX
The Big Crunch

When we throw a stone into the air and it reaches a certain height, the stone follows a parabolic arc while falling back to earth. According to the fundamental theory of charge, the projectile, whose mass is positive, is attracted to the Earth, whose mass is also positive, through a negative energy quantized by gravitons generated virtually by the quantum fluctuations of vacuum. By generalizing such a conception, we can infer that expanding 4-dimensional spacetime SU(22), whose electrical charge is positive of category 3 e^3, represents a projectile launched from 4-dimensional spacetime at rest SU(18), whose electrical charge is positive of category 1 e. Both will end up being attracted by the negative electrical charge -e of the gravitons SU(4), generating the contracting 4-dimensional spacetime SU(24), phenomenon known as the Big Crunch:

$$SU(24) = \frac{SU(22) \times SU(18)}{SU(4)} = \frac{e^3 \times e}{-e} = -e^3 \qquad (127)$$

Next, contracting 4-dimensional spacetime SU(24)-e^3 will be attracted by the cosmological constant SU(21) e^2 to give rise to an infinitely dense point SU(26):

$$SU(26) = \frac{SU(24)}{SU(21)} = \frac{-e^3}{e^2} = -e \qquad (128)$$

Decay of the 3 atomic components at high energies

Within the infinitely dense point of equation (128), the 3 atomic components, the levorotatory electron, the atomic proton and the neutron, will decay through the fifth messenger particle of the weak interactions, the Goldstone boson, which according to equations (63) and (64), has a positive electrical charge of category 2 e^2, spin 0 and mass 0. This final property determines that its scope is infinite.

Decay of the levorotatory electron at high energies

The levorotatory electron, through the Goldstone boson e^2, decays to a negative electrical charge of category 2 $-e^2$, a neutrino $-D$ and a positive magnetic charge D:

$$\text{Levorotatory electron decayed} = \frac{e^2 \text{ X } -e^2 \text{ X } -D}{D} = \frac{e^2 \text{ X } -e^2 \text{ X } \cancel{D}}{\cancel{D}}$$
$$= e^2 \text{ X } -e^2 = +1 \text{ X } -1 = -1 = -e^2 \quad (129)$$

Spin$+1/2$ of the levorotatory electron = Spin 0 of the $\qquad$ (130)
Goldstone boson + spin$+1/2$ of the neutrino

Decay of the atomic proton at high energies

The atomic proton, through the Goldstone boson e^2, decays into a levorotatory electron $-e^2$ and a dextrorotatory electron $-e^2$:

Atomic protons decayed$= e^2 \text{ X } -e^2 \text{ X } -e^2 = +1 \text{ X } -1 \text{ X } -1 = +1 = e^2$ (131)

Atomic proton spin 0=Goldstone boson spin 0 + Spin $\qquad$ (132)
$+1/2$ of levorotatory electron + spin $-1/2$ of dextrorotatory electron

Neutron decay at high energies:

The neutron, through the Goldstone boson e^2, decays into a levorotatory positron e2, a dextrorotatory positron e^2 and an antineutrino D:

Neutron decayed $= \dfrac{e^2 \text{ X } e^2 \text{ X } e^2}{D} = \dfrac{+1 \text{ X } +1 +1}{+1}$

$\qquad = \dfrac{+1}{+1 \text{ X the constant-}1} = \dfrac{+1}{-1} = \dfrac{e^2}{-e}$

$\qquad = -e$ \hfill (133)

Neutron Spin$+1/2$ = Goldstone boson Spin 0 + Levorotatory (134)
positron spin$+1/2$ + Dextrorotatory positron spin-$1/2$ +
Antineutrino spin$+1/2$

According to equations (129), (131) and (133), the three atomic components, when decaying at high energies related to the Big Crunch, will give rise to the six fundamental leptons: the neutrino, the levorotatory electron, the dextrorotatory electron, the levorotatory positron, the dextrorotatory positron and the antineutrino, which constitute the basic components from which the Universe will begin a new cycle.

Black holes

Within every black hole, whose energy is as high as those of the Big Crunch, processes similar to those described by the formulas (129), (131) and (133) are produced, resulting in the three atomic components, through the Goldstone boson, decaying into the six fundamental leptons, a negative electrical charge of category 2 and a positive magnetic charge, which, according to the following equation, constitute an infinitely dense mass M-:

$$M^- = \text{Levorotatory Electron decayed } \frac{(e^2\,X\,-e^2\,X\,-D}{D)}$$

$$\frac{\text{Atomic proton decayed } (e^2\,X\,-e^2\,X\,-e^2)}{\text{Decayed neutron } \frac{(e^2\,X\,e^2\,X\,e^2}{D)}}$$

$$= \frac{\dfrac{=e^2\,X\,-e^2\,X\,-D}{D}\;X\;\dfrac{e^2\,X\,e^2\,X\,e^2}{D}}{e^2\,X\,-e^2\,X\,-e^2}$$

$$= \frac{\dfrac{e^2\,X\,\!-\!e^2\,X\,\!-\!\cancel{D}}{\cancel{D}}\;X\;\dfrac{e^2\,X\,e^2\,X\,e^2}{D}}{e^2\,X\,-e^2\,X\,-e^2}$$

$$= \frac{e^2\,X\,e^2}{D} = \frac{+1\,X\,+1}{+1} = \frac{+1}{+1\,X\text{ the constant }-1} = \frac{+1}{-1}$$

$$= \frac{e^2}{-e} = -e \tag{135}$$

Spin +1 of M-=Goldstone boson spin 0 + spin+1/2 of (136) neutrino + levorotatory electron spin +1/2 + dextrorotatory electron spin -1/2 + levorotatory positron spin +1/2 + spin -1/2 of dextrorotatory positron + spin+1/2 of antineutrino

According to equations (135) and (136), the atom, whose electrical charge is positive in category 1 and whose spin is +1, when disintegrating at high energies and becoming an infinitely dense mass, which constitutes the fifth phase of matter, retains its spin, but reverses the sign of its electrical charge.

Inside a black hole, the infinitely dense mass -e is violently attracted by the inflaton e3 to generate x rays $-e^4$:

$-e^4 = -e \times e^3$ (137)

X rays, due to their negative electrical charge, tend to be repelled by the infinitely dense mass of the black hole, causing the phenomenon known as Hawking's radiation. In contrast, ultraviolet light, gamma rays and visible light, which have positive electrical charges, are attracted to this mass and are unable to escape. Based on these approaches, we can better understand another phenomenon called Einstein's gravitational lens: while the Sun or a galaxy, because of its positive mass, is capable of deflecting visible light from a distant star, a black hole, due to its negative mass, absorbs it, which demonstrates that, at a fundamental level, gravitational interactions, more than a geometric phenomenon, constitute an electromagnetic manifestation through which electrically charged objects interact through electrically charged gravitons.

Hawking Radiation

X rays $-e^4$ are repelled by the infinitely dense mass -e of a black hole by means of gravitons -e, generating a negative gravitational charge $-e^4$:

$$-e^4 = \frac{-e^4 \times -e}{-e}$$ (138)

Infrared light -e is repelled by the infinitely dense mass -e of a black hole by means of gravitons -e, generating a negative gravitational charge -e:

$$-e = \frac{-e \times -e}{-e}$$ (139)

Einstein's gravitational lens

On its trajectory towards the observer, visible light e from a distant star is deflected by the Sun e through gravitons -e, generating a negative gravitational charge -e:

$$-e = \frac{e \times e}{-e} \qquad (140)$$

The sunlight e is reflected by the Moon e through gravitons -e, generating a negative gravitational charge -e:

$$-e = \frac{e \times e}{-e} \qquad (141)$$

Reverse gravitational lens

In its trajectory towards the observer, visible light from a distant star is absorbed by the infinitely dense mass -e of a black hole by means of gravitons -e, generating a positive gravitational charge e:

$$e = \frac{e \times -e}{-e} \qquad (142)$$

In its trajectory towards the observer, x-rays $-e^4$ from a black hole are absorbed by the Sun e through gravitons -e, generating a positive gravitational charge e^4:

$$e^4 = \frac{-e^4 \times e}{-e} \qquad (143)$$

Experimental support

Experimental observations can show that stars such as the Sun, due to their positive electrical charge of category 1, let out ultraviolet light, gamma rays, visible light and radio waves that they produce in their nuclei, but absorb X rays and infrared light. The exact opposite is true of black holes, in accordance with the approaches of the fundamental theory of charge.

CHAPTER SEVEN
The experimental consequences of the new theory

The approaches of the fundamental theory of charge generate serious experimental consequences, as can be seen below.

A. There are 4 categories of electric charge: the neutrino, due to its positive electric charge of category 1, usually accompanies the electron, of negative electric charge of category 2. On the other hand, within the solar core, masses endowed with a positive electrical charge of category 1 are violently repelled by the positive electrical charge of category 3 of the speed of light to produce gamma rays, which, due to their positive electrical charge of category 4, can escape from the Sun whose electrical charge is positive of category 1.

B. The magnetic monopole is equivalent to the electric charge of category 1: according to equation (51), the neutron has a positive magnetic charge, which manifests itself as a negative electric charge of category 1, which was detected experimentally at the beginning of the 20th century, during the work on atomic theory. On the other hand, the atom, due to its negative magnetic charge which becomes a positive category 1 electrical charge, can be bombarded in the laboratory, not by protons, but by neutrons.

C. Quarks, in addition to fractional electrical charges, have fractional magnetic charges: the 3 components of the neutron udd, having fractional magnetic charges -1/3, 2/3 and 2/3 respectively, which add up to +1, explain why this baryon is not, as is believed, electrically neutral, but rather possesses a negative electrical charge of category 1, which in turn causes the atom not to be electrically neutral but rather to carry a positive electrical charge of category 1.

D. There are 7 fundamental interactions: the first 5, electro-magnetism, weak interactions, strong interactions, quantum gravity and supra-force, function on the basis of a total of 9 messenger particles while the other 2, the constant speed of light and the cosmological constant, operate through two messenger charges. In effect, the virtual photon favors the attraction between a proton and an electron (electromag-netism); at relatively high energies, for example, within a proton, the flavor uses the particles w^+ and z to become d and thus generates a neutron in which the flavor d, through the particles w⁻ and the Higgs boson, is transformed into u to create a proton (weak interactions); at high energies, inside a black hole, for example, the fifth messenger particle of weak interactions, the Goldstone boson, causes the decay of the 3 atomic components in order to generate an infinitely dense mass; inside the atomic nucleus, gluons, thanks to their pos-itive electrical charge of category 3, cause protons and neu-trons to unite (strong interactions); the positive masses of the Sun and the Earth, through the negative energy of gravitons, attract each other (quantum gravity); within the solar core, the inflaton, which has a positive electrical charge of catego-ry 3, violently repels positive masses to generate ultraviolet light (super force); likewise, within the solar core, the con-stant speed of light, provided with a similar electrical charge and using the same procedure, gives rise to gamma rays; finally, the cosmological constant, by means of a positive category 2 electrical charge, causes resting 4-dimensional spacetime, whose electrical charge is positive category 1, to generate the expanding 4-dimensional spacetime, which causes galaxies, as Hubble discovered, to move away from each other.

E. There are 6 fundamental leptons: they can be detected experimentally within the quantum fluctuations of vacuum where they form 3 virtual pairs of particles and antiparticles, generating the messenger particles. In fact, according to the fundamental theory of charge, a virtual photon is the product of the interaction between one of these virtual pairs: a levorotatory electron and a levorotatory positron, in the presence of 2 opposing magnetic charges; the particle w⁻, created virtually by a levorotatory electron and a levorotatory positron, in the presence of a positive category 2 electrical charge; the particle w⁺, by a dextrorotatory positron and a dextrorotatory electron, in the presence of a category 2 negative electrical charge; the particle z, by a levorotatory electron and a dextrorotatory positron, by a negative magnetic charge; the Higgs boson, by the virtual unification of the particles w⁻, w⁺ and z; the Goldstone boson, by the virtual unification of these particles with the Higgs boson ; the gluon, by a levorotatory electron and an antineutrino, in the presence of a positive electrical charge of category 2; the graviton, by a levorotatory electron, a levorotatory positron, a neutrino and an antineutrino; Finally, the inflaton, by a dextrorotatory electron, a levorotatory positron, a dextrorotatory positron and an antineutrino, emphasizing that all these data points were obtained with mathematical consistencies and in observance of the only two exact symmetries of Nature: the laws of charge and angular momentum.

Suggested experiments

In order to verify the validity of the fundamental theory of charge, the following experiments are suggested to international laboratories:

A. Interactions between neutrons, electrons and positrons are promoted. Predictions: Neutrons, due to their negative electrical charge of category 1, will repel electrons, but will attract positrons.

B. The chirality of both proton and neutron is observed. Predictions: the proton, being made up of a triplet of quarks that add up to -1/2 spin, has right chirality; that is to say, it is dextrorotatory (turns to the right). On the other hand, the neutron, made up of a triplet of quarks that add up to spin +1/2, has left chirality; that is to say, it is levorotatory (turns to the left).

C. Equation (98) is used to describe the gravitational interactions between the Sun and each of its planets. Predictions: the planetary orbits will be explained, including the precession in the perihelion of Mercury.

D. The interior of the atomic nucleus is scrutinized. Predictions: a powerful negative electrical charge will be detected, which according to equation (91), keeps the atomic nucleus cohesive, preventing electrically positive particles such as protons and gluons from escaping, but allowing the eventual release of other electrically negative particles such as neutrons.

E. The chirality of the 6 fundamental leptons is observed. Predictions: the neutrino, the antineutrino, the levorotatory electron and the levorotatory positron have left chirality and therefore their respective spins are positive while the dextrorotatory electron and the dextrorotatory positron have right chirality and therefore their spins are negative. Likewise, the non-existence of dextrorotatory neutrinos and antineutrinos will be ratified.

F. The chirality of the particles w^- and w^+ are observed. Predictions: it will be determined that w^- is levorotatory and w^+, dextrorotatory.

G. Instead of neutrons, protons are used to try to bombard atoms. Prediction: Such an operation will not be possible because both the proton and the atom are electrically positive.

H. The Sun is investigated in detail. Predictions: The king star, being electrically positive like the atom, attracts and retains electrically negative photons such as X rays and infrared light, allowing ultraviolet light, gamma rays, visible light and radio, which are electrically positive, to escape from its surface.

I. The decay of the free neutron is observed. Prediction: it will be determined that the negative beta disintegration does not involve particle w‾, but rather the Higgs boson which is also endowed with a negative electrical charge, although of category 1.

Understanding our environment

Through the fundamental theory of charge, it is possible to achieve a better understanding of the world around us, getting accurate explanations to questions related to the functioning of Nature.

Why do objects fall to earth?

The positive masses of the Earth and any other object attract each other by means of a negative energy carried by gravitons, which are virtually generated by the quantum fluctuations of vacuum. The apple that Newton saw falling, had a positive mass that was intercepted by the negative energy of the gravitons and thrown into the positive mass of the Earth.

Why do we live in a 4-dimensional spacetime?

According to equation (118), our 4-dimensional spacetime is the product of attraction, through gravitons, whose spin is +2, between the 2 halves of the Universe, dark matter, whose spin is +1, and positive matter, whose spin is +1.

The spin +4 that sums up the 3 factors is translated at the macroscopic level into the 4 dimensions of space time, which implies that the concept of spin of the microscopic world is manifested in the macroscopic world as the notion of dimension.

Why does the moon shine?

Our natural satellite inherits the positive electrical charge of category 1 from the atom. Consequently, it tends to reflect sunlight, which is electrically positive. If the Moon were electrically negative, it would retain such light and be an opaque body in the eyes of the observer, which is precisely what happens with dark matter and black holes.

Why, if the neutron is electrically negative, is a neutron star not dark and instead is rather bright according to experimental observations?

In fact, a neutron star, which is the product of the collapse of a star with less than 3 solar masses, is not only composed of these baryons but also of electrons and protons, which, according to equation (96), generate a positive category 1 electrical charge. When the mass of a collapsed star is greater than 3 solar masses, the 3 components of its atoms decay through the Goldstone boson, as described in equations (129), (131) and (133), to become the infinitely dense mass of a black hole. Neutrons, because they have half integer spin, cannot occupy the same quantum state due to the Pauli exclusion principle, which is why a neutron star cannot continue to contract. However, the infinitely dense masses, according to equation (136), possess integer spin, which allows them to occupy the same quantum state within a black hole.

Why, while gamma rays and ultraviolet light damage our skin, do X rays, which are also high-energy photons, penetrate through it without damaging it, generating radiographic plates?

Our skin, which is electrically positive, attracts and absorbs X rays, which are electrically negative, but is violently repelled by gamma rays and ultraviolet light, which are electrically positive, causing skin damage.

Why has the magnetic monopole not been observed experimentally?

The scientific community, without realizing it, has already observed the magnetic monopole, which always manifests itself as a category 1 electrical charge. When, at the beginning of the 20th century, at the height of the atomic theory, a weak negative electrical charge was detected on the surface of the neutron, at the bottom, a positive magnetic charge was being observed; when in experiments a neutrino, due to its positive electrical charge of category 1, is attracted by an electron, producing the phenomenon known as an electronic neutrino, the charge is nothing but the manifestation of a magnetic monopole.

Why, according to the fundamental theory of charge, do the so-called 6 fundamental leptons constitute the basic components of matter?

These leptons, in addition to combining to create each other, which in the quantum world is feasible, generate quarks, messenger particles, the 3 internal symmetries (electromagnetism, weak interactions and strong interactions), as well as quantum fluctuations of vacuum where precisely 3 virtual pairs of particles and antiparticles are created and destroyed continuously.

Finally, evidencing that at the precise moment of the Big Bang there was no excess of matter over antimatter and vice versa, the virtual levorotatory positron- dextrorotatory electron pairs, as well as the levorotatory electron- dextrorotatory positron, due to their opposite charges and spins, annihilated each other to become pure energy. However, the virtual neutrino-antineutrino pair, by having identical spins, although opposite charges, by annihilating each other, were transformed, not into pure energy, but into a positive mass, the precursor of positive or observable matter.

Why is it now possible, in the light of the fundamental theory of charge, to speak coherently about a time before the Big Bang?

The new laws of physics, unlike the current ones, attempt to explain with mathematical certainty and in accordance with the principles of symmetry the events prior to the Big Bang. In fact, according to equations (5-7), in an infinitely dense point, 3 virtual pairs of particles-antiparticles (the 6 fundamental leptons), through the symmetry of the charge, originated the 3 internal symmetries (electromagnetism, weak interactions and strong interactions) which, in turn, generated 7 external symmetries: quantum gravity SU(4), supra-force SU(5), quantum fluctuations of vacuum SU(6), infinitely dense mass SU(8), 10-dimensional space SU(10), 11-dimensional spacetime SU(11) and the constant speed of light Su(12) which, in turn, when the Big Bang occurred, gave rise to 7 other external symmetries: primary X rays SU(13), primary gamma rays SU(14), dark matter SU(16), observable matter SU(17), resting 4-dimensional spacetime SU(18), the cosmological constant SU(21) and expanding 4-dimensional spacetime SU(22). All the above symmetries have one thing in common: the charge factor, which allows them to be attracted or repelled to transform or be transformed.

Concluding remarks

Nature is an open book whose codes can be deciphered through a unique key. Current physics is swimming against the tide by trying to explain fundamental interactions based on different concepts, which are but different manifestations of the same phenomenon. The notion of field is used to describe electromagnetism, the notion of spontaneous symmetry rupture in the case of weak interactions, the notion of color in the case of strong interactions, and the notion of spatial-temporal curvature in the case of gravity. As a result, problems arise such as the calculation of strong interactions, the quantization of gravity and the unification of the 4 fundamental interactions into a supra-force.

The fundamental theory of charge, following the dictates of Nature, interprets its codes by means of a single concept, which is that of charge. Thus, the messenger particles of electromagnetism, weak interactions, strong interactions and quantum gravity have a common denominator: an electrical charge, which can be of category 1, 2 or 3. Such an approach not only makes it possible to explain these four fundamental interactions in a coherent way, but also gives us a better perspective, allowing us to determine the existence of three others: the supra-force, whose messenger particle, the inflaton, transforms the positive masses into ultraviolet light within the solar nucleus; the constant speed of light, whose messenger charge generates gamma rays within it; and the cosmological constant, whose messenger charge drives 4-dimensional spacetime to produce an expanding Universe.

The fundamental theory of charge, besides opening a fruitful field of work for experimental physicists who will have the mission of testing its predictions, places in the hands of theoretical physicists new tools and above all reconciles them with the true essence of Nature: an unique notion to interpret the different manifestations of the same phenomenon.

BIBLIOGRAPHY

BARROW, John; *Theories of Everything*, Clarendon Press, Oxford, 1991.

BARROW, John; *The Origin of the Universe*. Phoenix, Orion Books, London, 1994.

BARROW, John and SILK, Joseph; *The Left Hand of Creation*, Oxford University Press, 1994.

BOHNER, Gerhard; *The Early Universe*, Springer Verlag, 1988.

DAVIES, Paul; *God and The New Physics*, Penguin Books, England, 1990.

DAVIES, Paul; *The Mind of God*, Simon and Schuster, 1992.

GELL-MANN; *The Quark and The Jaguar*, W. H. Freeman and Company, New York, 1994.

GUTH, Alan; *The inflationary Universe*, Addison-Wesley, Helix Books, 1997.

HAWKING, Stephen; *A Brief History of Time*, Bantam Books, New York, 1988.

LOZANO LEYVA, Manuel; *El Cosmos en la Palma de la Mano*, Random House Mondadori, S.A, 2002.

WEINBERG, Steven; *Los Tres Primeros Minutos del Universo*, Alianza Editorial, 1978.